An OCEAN FOOD WEB

BY CARI MEISTER · ILLUSTRATED BY HOWARD GRAY

AMICUS ILLUSTRATED and **AMICUS INK**
are published by Amicus
P.O. Box 1329, Mankato, MN 56002
www.amicuspublishing.us

COPYRIGHT © 2021 Amicus. International copyright reserved in all countries. No part of this book may be reproduced in any form without written permission from the publisher.

Editor: Alissa Thielges
Designer: Kathleen Petelinsek

Library of Congress Cataloging-in-Publication Data
Names: Meister, Cari, author. | Gray, Howard (Howard Willem Ian), illustrator.
Title: An ocean food web / by Cari Meister ; illustrated by Howard Gray.
Description: Mankato, MN : Amicus, [2021] | Series: Ecosystem food webs | Includes bibliographical references. | Audience: Ages 6-9 | Audience: Grades 2-3 | Summary: "An illustrated narrative nonfiction journey to the Pacific that shows elementary readers how animals and plants in an ocean ecosystem survive in an interconnected food web"— Provided by publisher.
Identifiers: LCCN 2019037176 (print) | LCCN 2019037177 (ebook) | ISBN 9781645490050 (library binding) | ISBN 9781681526478 (paperback) | ISBN 9781645490852 (pdf)
Subjects: LCSH: Marine ecology—Pacific Ocean—Juvenile literature. | Food chains (Ecology)—Pacific Ocean—Juvenile literature.
Classification: LCC QH95 .M45 2021 (print) | LCC QH95 (ebook) | DDC 577.7—dc23
LC record available at https://lccn.loc.gov/2019037176
LC ebook record available at https://lccn.loc.gov/2019037177

Printed in the United States of America.

HC 10 9 8 7 6 5 4 3 2 1
PB 10 9 8 7 6 5 4 3 2 1

About the Author
Cari Meister has written more than 200 books for children, including the TINY series (Viking), and the FAIRY HILL series (Scholastic). She lives in the mountains of Colorado with her husband and four sons. Cari loves to visit schools and libraries. Find out more at carimeister.com.

About the Illustrator
Howard Gray has illustrated a selection of fiction and non-fiction children's books. He has always considered himself an artist, but with a PhD in dolphin genetics, he has a background in zoology. He is now pursuing his dream career in children's illustration from the picturesque city of Durham, UK. Find out more at www.howardgrayillustrations.com.

Wow, the Pacific Ocean is big! It is the largest ocean in the world. From the surface, it looks calm. Underneath, these open waters are full of life. Thousands of marine animals live here. They make up a food web, which shows what animals eat to get energy. Let's dive in!

Animals of all sizes call the ocean home. You'll find gigantic blue whales eating the tiniest krill. All the animals depend on one another to survive.

Living things need energy to live. Animals get energy by eating. Plants get energy from the sun.

Phytoplankton drift along in the ocean water. These tiny plants are so small you'd need a microscope to see them. Tiny krill have no trouble finding them, though. Phytoplankton are the first group in the food web: producers. They make their own food.

phytoplankton enlarged

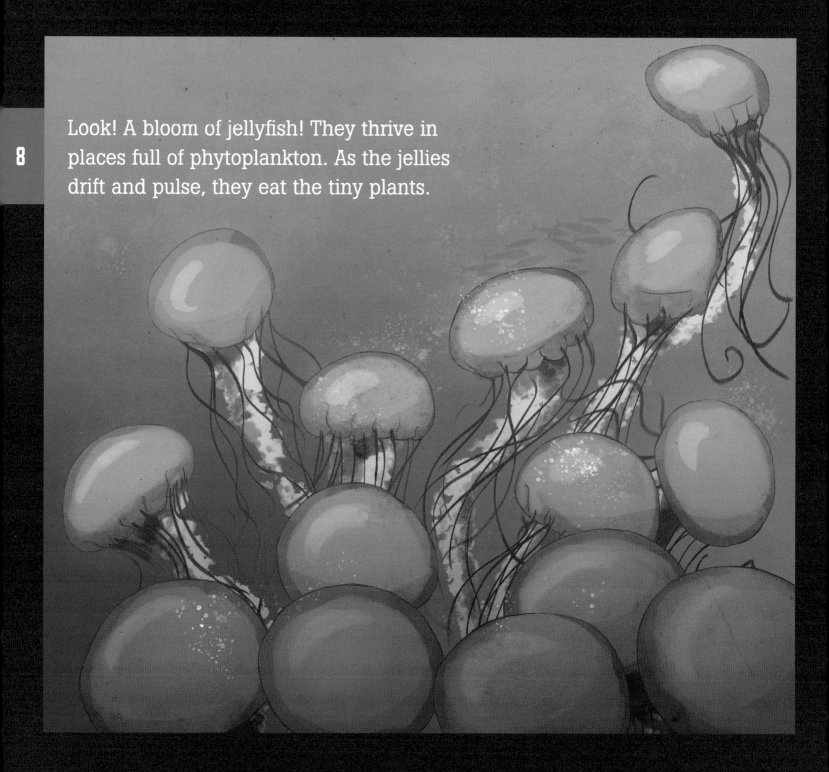

Look! A bloom of jellyfish! They thrive in places full of phytoplankton. As the jellies drift and pulse, they eat the tiny plants.

Snails, krill, plankton and many other animals eat phytoplankton. Plant-eating animals are called herbivores. They are primary consumers. The open ocean is full of them.

Jellyfish are also a popular meal in the ocean. Fish, such as herring and mackerel, eat them. So do leatherback sea turtles. These predators are secondary consumers.

Predators are important to an ecosystem. Without them, there would be too many herbivores and not enough plants to go around.

Whoa! Do you see that? It's a giant school of mackerel. There are thousands of fish in there! They sense something. What's going on?

13

A mako shark! An adult mako is an apex predator. Other animals don't usually hunt it.

The mackerel swim faster. They try to hide behind each other. This makes a tight bait ball. Will it scare the shark?

15

Not this time! The shark pushes through the spinning mass. Chomp, chomp! Tuna join the frenzy. They eat their fill. Birds flying over notice the action. They dip down and grab some fish.

Another shark swims by. The mackerel have left, but that's okay. Mako sharks are like most animals in the ocean. They feed on different kinds of animals. This shark dines on a small squid tonight.

When an animal dies, scavengers, like king crabs, quickly eat its dead body. Decomposers, like marine worms and bacteria, break down whatever is left.

The nutrients go back into the ecosystem. Plants use them to make food. Then the whole process starts again!

An Ocean Food Web

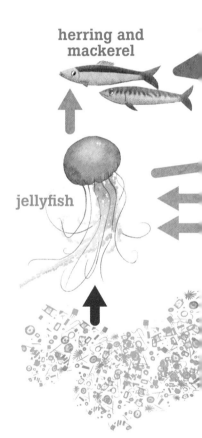

apex predator An animal that has no natural predators; it is the top consumer in a food web.

secondary consumer An animal that can eat both plants and other animals, but also gets eaten by other animals.

primary consumer An animal that eats only plants and is eaten by other animals.

producer A plant that makes its own food.

scavengers and **decomposers** Animals that break down dead animals and plants by eating them, which helps nutrients go back into the ground.

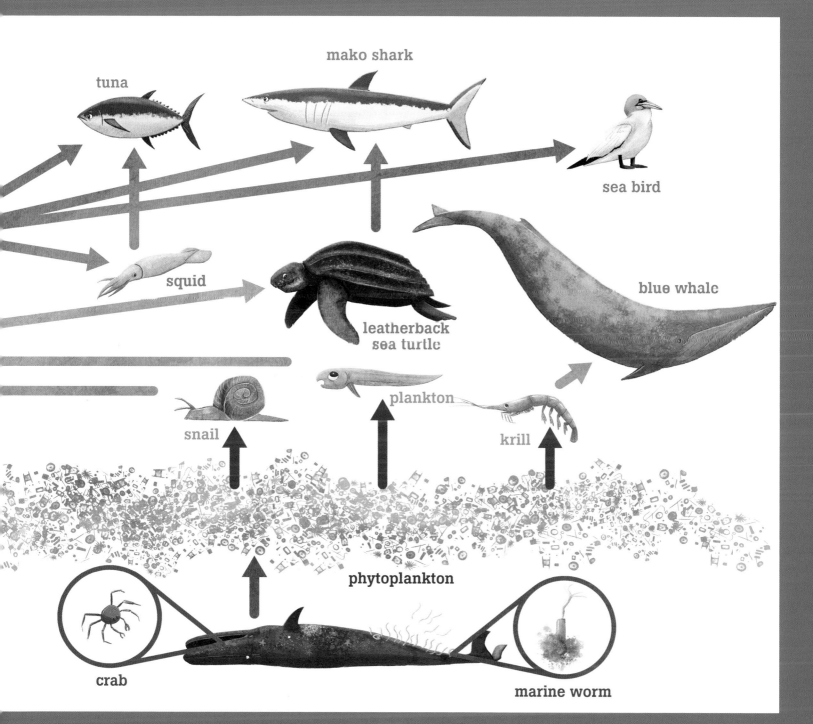

GLOSSARY

bacteria Microscopic living things that live in soil and water or in the bodies of plants and animals.

bloom A large group of jellyfish that gather together.

ecosystem A community of plants and animals that live in a certain area.

food web A system of how animals and plants in an ecosystem relate; it shows who eats what.

WEBSITES

Cool Classroom | Food Web Game
http://coolclassroom.org/classic/cool_windows/home.html

Ecosystems: Climate Kids | 10 Interesting Things About Ecosystems
https://climatekids.nasa.gov/10-things-ecosystems/

National Geographic Kids | Ocean Habitat
https://kids.nationalgeographic.com/explore/nature/habitats/ocean/

Every effort has been made to ensure that these websites are appropriate for children. However, because of the nature of the Internet, it is impossible to guarantee that these sites will remain active indefinitely or that their contents will not be altered.

READ MORE

Callery, Sean. **Ocean**. Life Cycles. New York: Kingfisher, 2018.

Jacobson, Bray. **Food Chains and Webs**. New York: Gareth Stevens Publishing, 2020.

Pettiford, Rebecca. **Ocean Food Chains**. Minneapolis: Jump!, 2016.